Syntropy Precognition and Retrocausality

Ulisse Di Corpo

www.sintropia.it

Copyright © August 2017 Ulisse Di Corpo

ISBN: 9781549522451

CONTENTS

Prologue	1
Syntropy, retrocausality and precognition	3
How to enhance precognitions, intuitions and wealth	35
Retrocausality: changing the past	57
Epilogue	75

PROLOGUE

On the 30th of July 2017 I received the following email from Marty Rosenblatt:

Hi Ulisse, ..., Are you interested in focusing on Precognition (forward looking) and Retrocausality (backward looking), which of course do relate directly to Syntropy and time? I was thinking of the following talk: Syntropy, Precognition and Retrocausality".

Syntropy implies a change of paradigm, according to which life is finalized and guided by attractors. This change of paradigm merges qualitative with quantitative, objective with subjective, visible with invisible, opening the way to retrocausality and precognition in science.

The applications and implications touch all the fields of life and can have great consequences, based on a concept of time and causality which is at first counterintuitive. However, it is simple, has important implications and it is accessible to all, scientists and non-scientists.

SYNTROPY, RETROCAUSALITY AND PRECOGNITION

The notion of energy comes from the fact that physical systems possess a quantity that can be turned into a force. Despite the fact that it is used and studied Feynman noted that *"it is important to realize that in physics today we have no knowledge of what energy is."*[1]

The energy-mass relation $E=mc^2$ that we all associate with Einstein, was first published by Oliver Heaviside in 1890[2], then by Henri Poincaré in 1900[3] and by Olinto De Pretto in 1904[4]. Olinto De Pretto presented it at the *Reale Istituto Veneto di Scienze* in an essay with a preface by the astronomer and senator Giovanni Schiaparelli. It seems that this equation reached Einstein through his father Hermann who was responsible for the lighting systems in Verona and who, as director of the *"Privilegiata Impresa Elettrica Einstein"*, had frequent contacts with the Fonderia De Pretto that produced the turbines for electricity.

---

[1] Feynman, R., *The Feynman Lectures on Physics*, vol. 1 chapter 4: http://www.feynmanlectures.caltech.edu/I_04.html
[2] Auffray, J.P., *Dual origin of E=mc²*: http://arxiv.org/pdf/physics/0608289.pdf
[3] Poincaré, H., Arch. néerland. sci. 2, 5, 252-278 (1900)
[4] De Pretto O., Lettere ed Arti, LXIII, II, 439-500 (1904), Reale Istituto Veneto di Scienze: www.cartesio-episteme.net/st/mem-depr-vf.htm

However, the $E=mc^2$ does not take into account the momentum, which is also a form of energy, and in 1905 Einstein added the momentum (p), thus obtaining the energy-momentum-mass equation ($E^2=m^2c^4+p^2c^2$). Since energy is squared (E^2) and in the momentum (p) there is time, a square root is used and there are two solutions: negative time energy and positive time energy. Positive time energy implies causality, whereas negative time energy implies retrocausality: the future that acts backwards into the past. This was considered impossible and to solve this paradox Einstein removed the momentum, given the fact that it is practically equal to zero compared to the speed of light (c). In this way, he returned to the $E=mc^2$.

Though, in 1924 the spin of the electrons was discovered. The spin is an angular momentum, a rotation of the electron on itself at a speed close to that of light. Since this speed is very high, the momentum cannot be considered equal to zero and in quantum mechanics the energy-momentum-mass equation must be used with its uncomfortable dual time solution.

The first equation that combined relativity and quantum mechanics was formulated in 1926 by Oskar Klein and Walter Gordon and has two solutions: advanced and delayed waves. Advanced waves were rejected, since they imply retrocausality which was considered impossible. The second equation, formulated in 1928 by Paul Dirac, also has two-time solutions: electrons and neg-electrons (now called positron). The existence of positrons was proved in 1932 by Carl Andersen.

However, retrocausality was considered unacceptable and the backward-in-time solution was declared impossible.

Luigi Fantappiè, born in Viterbo (Italy) on the 15th of September 1901, graduated in pure mathematics at the age of 21 at the Normale di Pisa, the most exclusive Italian University, and became full professor at the age of 27. He was well known and appreciated among physicists to the point that in 1951 Oppenheimer invited him to become a member of the Institute for Advanced Study in Princeton and work directly with Einstein.

As a mathematician Fantappiè could not accept that half of the solutions of the fundamental equations where rejected and in 1941, while listing the properties of the forward and backward in time energy, Fantappiè discovered that forward in time energy is governed by the law of entropy, whereas backward in time energy is governed by a complementary law that he named syntropy, combining the Greek words *syn* which means converging and *tropos* which means tendency.

Listing the mathematical properties of syntropy Fantappiè discovered: energy concentration, increase in differentiation, complexity and structures: the mysterious properties of life! In 1944 he published the book "*Principi di una Teoria Unitaria del Mondo Fisico e*

Biologico"[5] (Unitary Theory of the Physical and Biological World) in which he suggests that the physical-material world is governed by the law of entropy and causality, whereas the biological world is governed by the law of syntropy and retrocausality.

We cannot see the future and therefore retrocausality is invisible! The dual energy solution suggests the existence of a visible reality (causal and entropic) and an invisible reality (retrocausal and syntropic).

The first law of thermodynamics states that energy is a constant, a unity that cannot be created or destroyed but only transformed, and the energy-momentum-mass equation suggests that this unity has two components: entropy and syntropy. We can therefore write: *1=Entropy+Syntropy* which shows that syntropy is the complement of entropy.

Syntropy is often mistaken with negentropy. However, it is fundamentally different since negentropy does not take into account the direction of time, but considers time only in the classical way: flowing forward.

Life lies between these two components: one entropic and the other syntropic, one visible and the other invisible, and this can be portrayed using a seesaw with entropy and syntropy playing at the opposite sides, and life at the center.

---

[5] Fantappiè, L., *Principi di una teoria unitaria del mondo fisico e biologico*. Humanitas Nova, Roma 1944: www.amazon.it/dp/B07RYVS89S

This suggests that entropy and syntropy are constantly interacting and that all the manifestations of reality are dual: emitters and absorbers, particles and waves, matter and anti-matter, causality and retrocausality.

Fantappiè failed to provide experimental proof to his theory, since the experimental method requires the manipulation of causes before observing their effects. Recently, random event generators (REG) have become available. These systems allow to perform experiments in which causes are manipulated after their effects: in the future.

One of the first experimental studies on retrocausality, by Dean Radin[6] of IONS (Institute of

---

[6] Radin, D.I., Unconscious perception of future emotions: An experiment in presentiment, Journal of Scientific Exploration, 1997, 11(2): 163-180:
http://deanradin.com/articles/1997%20presentiment.pdf

Noetic Sciences), measured heart rate, skin conductance and blood pressure in subjects who were presented with blank images for 5 seconds followed by images that, based on a random event generator, could be neutral or emotional. The results showed a significant activation of the parameters of the autonomic nervous system, before the presentation of emotional images.

In 2003, Spottiswoode and May[7], of the Cognitive Science Laboratory, replicated this experiment by performing a series of controls to study possible artifacts and alternative explanations. The results confirmed those already obtained by Radin. Similar results were obtained by other authors, such as McCarthy, Atkinson and Bradley[8], Radin and Schlitz[9] and May, Paulinyi and Vassy[10], always using the parameters of the autonomic nervous system.

---

[7] Spottiswoode, P., and May, E., *Skin Conductance Prestimulus Response: Analyses, Artifacts and a Pilot Study*, Journal of Scientific Exploration, 2003, 17(4): 617-641:
pdfs.semanticscholar.org/4043/2bc0a6b83f717dca2349b189ebdcbe7b3df9.pdf

[8] McCarthy, R., Atkinson, M., and Bradley, R.T., *Electrophysiological Evidence of Intuition: Part 1*, Journal of Alternative and Complementary Medicine; 2004, 10(1): 133-143:
https://www.ncbi.nlm.nih.gov/pubmed/15025887

[9] Radin, D.I., and Schlitz, M.J., *Gut feelings, intuition, and emotions: An exploratory study*, Journal of Alternative and Complementary Medicine, 2005, 11(4): 85-91:
www.ncbi.nlm.nih.gov/pubmed/15750366

[10] May, E.C., Paulinyi, T., and Vassy, Z., *Anomalous Anticipatory Skin Conductance Response to Acoustic Stimuli: Experimental Results*

Daryl Bem[11], psychologist and professor at the Cornell University, describes nine well-established experiments in psychology conducted in the retrocausal mode in order to get the effects before rather than after the stimuli. For example, in a priming experiment, the subject is asked to judge whether the image is positive (pleasant) or negative (unpleasant) by pressing a button as quickly as possible. The reaction time is recorded. Just before the positive or negative image, a *prime* is presented briefly, below the perceptual threshold so that it is not perceivable at a conscious level. It has been observed that subjects tend to respond more quickly when the *prime* is congruent with the following image, whether it is a positive or a negative image, while the reactions become slower when they are not congruent, for example when the *prime* is positive while the image is negative.

In retro-priming experiments, the *prime* is shown after, rather than before the subject responds, based on the hypothesis that this "inverse" procedure can retrocausally influence the reaction time. The

---

and Speculation about a Mechanism, The Journal of Alternative and Complementary Medicine. August 2005, 11(4): 695-702: www.ncbi.nlm.nih.gov/pubmed/16131294

[11] Bem, D., *Feeling the future: Experimental evidence for anomalous retroactive influences on cognition and affect*, Journal of Personality and Social Psychology, 2011, 100(3): 407– 425, DOI: 10.1037/a0021524,
https://pdfs.semanticscholar.org/79ec/e4f787af713d82924e41d8c17ab130f4b22d.pdf

experiments were conducted on more than a thousand subjects and showed retrocausal effects with statistical significance of $p<1.34 \times 10^{11}$, a possibility on 134,000,000,000 of being mistaken when affirming the existence of the retrocausal effect.

Syntropy explains these results in the following way:

"Since life nourishes on syntropy, and syntropy flows backwards-in-time, the parameters of the autonomic nervous system that supports vital functions must react in advance to future stimuli."

This general hypothesis was translated in the following working hypothesis:

"Heart rate and skin conductance should react in advance to future stimuli."

Before starting the experiments, in late 2007, an assessment of heart rate and skin conductance measuring devices was carried out. One requirement was to associate the measurements with the time of the clock of the computer. The problems were the following:

- Measuring devices used a different clock from the one used by the computer.

- Data did not consider the time required to produce the measurements.
- Built-in software did not satisfy the synchronization of measurements with the stimuli presented on the PC monitor.
- Proprietary software did not allow to access directly the device.

In order to try to overcome these difficulties, a laboratory in North Italy provided some devices, but still it was impossible to establish a satisfactory synchronization between the measurements and the stimuli which were presented on the PC monitor.

In December 2007 the assessment was extended to devices used in the field of sports training. Most devices showed the following limits:

- Heart rate measurements were stored in a wrist watch, using a different clock from the one used by the PC.
- The information was stored without any compensation for the delay due to the measurement.
- Some devices showed errors in the measurements.

After a long evaluation, the "home training" device produced by SUUNTO was chosen. This system included a thorax belt for measuring heart rate parameters and a USB interface (PC-POD) which received data by radio signals (using digital formats which eliminated any possibility of interference) directly on the PC on which the experiment was carried out and using in this way the same clock of the PC.

The SUUNTO heart rate monitor device saved the heart frequency information every second, together with PC clock information (year, month, day, hour, minute and second). Data was saved compensating the delay due to the time necessary to perform the measurement and to process the information. The heart rate data could therefore be synchronized with the data saved by the PC software used in the presentation of stimuli.

The heart rate information was saved as an integer number, without any decimal values. The technical support unit of SUUNTO was contacted in Helsinki and gave full cooperation sending all the necessary documentation, software and ddl libraries. SUUNTO underlined that synchronization and precision of measurements are diverging parameters. A precise synchronization diminishes the precision of the measurements. An integer value of the heart rate, provided every second, can be considered an excellent measurement.

The SUUNTO "home training" device was developed in order to monitor sports training activities and can be used in the most extreme conditions, for example underwater. It does not require the use of gel in order to receive the heart signals and it is extremely simple to use. It does not require the presence of an assistant in the same room in which the experiment is carried out. The only limit was observed in cold weathers when the skin gets dry and this limits the possibility to measure the heart rate parameters.

Experimental trial

Several experimental designs were tested and a design divided in 3 phases was chosen:

— In the presentation phase: 4 stimuli were shown individually on the PC monitor and the heart rate was measured.
— In the choice phase: stimuli were shown together on the PC monitor and the subject had to guess which one the computer would select.
— In the random selection phase, the computer selected one of the 4 stimuli (target stimulus), using a random procedure, and showed it on the PC monitor, full screen.

The initial hypothesis was that in the event of retrocausality heart rate measurements in phase 1 (the presentation phase) would be significantly different among target images (those which will be selected randomly by the computer in phase 3).

The first tentative experiments used stimuli made of black bars placed horizontally, vertically and diagonally on a white background. Data analyses did not show any significant difference among heart rates measured in phase 1.

The hypothesis was therefore analyzed more in depth and it was found that the "syntropy theory" posits that retrocausality is mediated by feelings and, therefore, in order to assess differences in heart rates measured in phase 1, stimuli in phase 3 should arise feelings. Following this indication, it was decided to use 4 elementary colors: blue, green, red and yellow. Using colors, a strong difference in heart rate frequencies in phase 1 was observed in correlation to the target shown in phase 3.

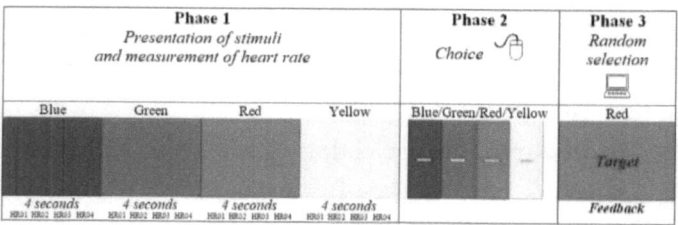

The experimental trial was the following:

— *Phase 1*, in which 4 colors were displayed one after the other on the computer screen and the heart rate was measured.

- *Phase 2*, in which an image with 4 colored bars was displayed and the subject had to try to guess the color that the computer would have selected.
- *Phase 3*, in which the computer randomly selected the target color and showed it full screen.

This experimental trial was repeated 100 times for each subject.

Hypothesis

The hypothesis was that in the case of a retrocausal effect, differences should be observed among the heart rates measured in phase 1 in correlation with the target color selected in phase 3 by the computer.

Random Events Generator

In order to test retrocausal effects it is necessary to use Random Event Generators.

In a random sequence each term is totally independent from the previous and following terms, no rule links different parts of the sequence. This condition is known as unpredictability of random sequences and it is referred to as "lack of memory": the process of random selection does not recall any information about the values which were selected previously, and cannot be used for the prediction of the values which will be selected in the future.

Random sequences imply:

– *Unpredictability*. The knowledge of any portion of the random sequence does not provide useful information in order to predict the outcome of any other element of the sequence. In other words, the knowledge of the first k values does not provide any element in order to predict the value k+1: this property is called unpredictability.
– *Equiprobability*. A sequence is random if in each position each value has the same probability to be selected. In the case of a dice, each side has the same probability to be selected. Similarly, equal probability is expected when using a coin: during each tossing, heads and tails have the same probability to show. Equiprobability implies independent sequences as it requires that the outcome of each selection is independent from any previous selection.
– *Irregularity*. Unpredictability requires random sequences to be irregular and not repetitive.
– *Absence of order*. In random sequences no type of structure or order can be detected.

The basic difference between causal and random can be traced back to the fact that causal events can be predicted, whereas random events cannot be predicted. A random sequence can be defined as a sequence that no cognitive process will ever be able to predict.

Pseudorandom and random

Computer languages usually use the word *random* to identify the instruction which starts the algorithm for the random selection of numbers. In the experiments described in this paper the Delphi-Pascal programming language was used. Delphi-Pascal has a predefined random sequence of 2^{32} numbers, which can be assessed through a pointer which can be defined by the user or by the value of the built-in clock. Delphi-Pascal uses the following instructions:

– *Randomize* reads the value of the built-in clock and uses this value as the pointer to the predefined random sequence;
– *Random* reads the value of the predefined sequence using the pointer selected by the randomize instruction.

The user can also define a personalized pointer. This option is generally used to encrypt information. Using the same pointer, the random sequence will always be the same.

Random sequences produced by computers are named *pseudorandom* since loops require always the same processing time, and the new random value will therefore be determined by the previous one. The problem with random sequences generated by computers arises when the randomize procedure is recalled in a loop, since random numbers will be determined by the first value which was selected: the first value determines the second one, and so on, and the condition of independency between different terms is lost.

Usually the fact that computers produce pseudorandom sequences is considered insignificant. However, in experiments which want to test retrocausality, and which are based on the assumption of unpredictability, a pseudo-random sequence would inevitably be an artifact.

Luckily the solution is relatively simple. The problem arises from the fact that the period of the loops is always the same. In order to overcome this problem, obtaining in this way pure random sequences, it is necessary to use loops which are based on unpredictable periods of time. This condition can be easily met when an external, unpredictable factor, is inserted in the loop and modifies its execution time.

In the experiments described in this paper subjects were asked to guess in phase 2 the color that the computer will select, pressing a button: the reaction time of the subject is unpredictable. In this way, the unpredictable reaction time of the subject makes loops time become unpredictable, and the value read from the built-in clock of the computer becomes independent from the other values, the independence is restored and the sequence becomes totally unpredictable: perfectly random.

Data Analysis

In this chapter only the fourth experiment will be discussed. A complete review of the experiments is available in: "*Retrocausality: experiments and theory*"[12] and "*A syntropic model of consciousness*"[13].

Since each subject completed 100 trials, data analysis could be performed for each single subject and results generally were strong from a quantitative and statistical point of view and showed the retrocausal effects in phase 1 associated to all the target colors (the colors chosen by the computer in phase 3).

However, this effect could go in different directions. In some subjects when red was the target color the heart rate increased in phase 1, whereas in other subjects it decreased. Considering all the

---

[12] Vannini, A. and Di Corpo, U., *Retrocausality: experiments and theory*, ISBN: 9781520275956, www.amazon.com/dp/1520275951
[13] Vannini, A., *A syntropic model of consciousness*, ISBN: 9781520834412, https://www.amazon.com/dp/1520834411

subjects together, these opposite effects were cancelling each other reducing the general effect or showing it only on some colors. For this reason, we started using for each subject "feed-back" tables where each line is relative to one of the 16 heart rates measurements in phase 1.

	Blue	Green	Red	Yellow
HR 1:	-0.671	2.200	-0.840	-1.103
HR 2:	-0.772	2.399	-0.556	-1.471
HR 3:	-0.950	2.467	-0.056	-1.766
HR 4:	-1.353	2.310	1.080	-2.054
HR 5:	-1.928	2.204	1.894	-1.892
HR 6:	-1.954	1.897	2.474	-1.993
HR 7:	-1.982	1.535	2.752	-1.755
HR 8:	-2.015	1.543	2.733	-1.704
HR 9:	-1.831	1.397	2.665	-1.704
HR 10:	-1.770	1.508	2.407	-1.691
HR 11:	-1.482	1.468	1.981	-1.641
HR 12:	-1.458	1.853	1.404	-1.637
HR 13:	-1.572	2.154	1.199	-1.679
HR 14:	-1.544	2.079	1.260	-1.676
HR 15:	-1.452	1.994	1.226	-1.661
HR 16:	-1.311	1.727	1.255	-1.541

Feed-back table of the retrocausal effect for one subject.

Since for each heart rate (HR) we have 100 values (100 trials), the difference between the mean value when the color was target and not target could be calculated.

The feed-back table was then graphically represented.

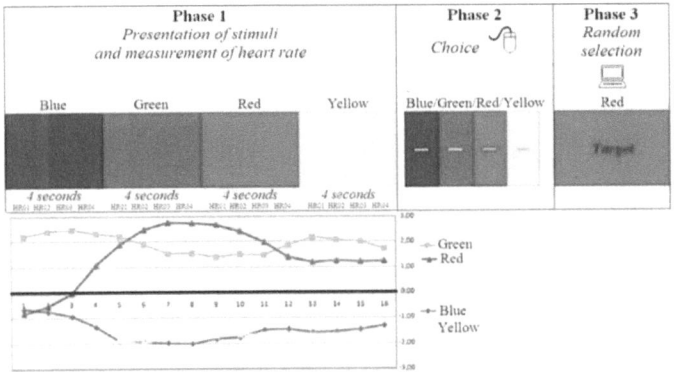

Graphical representation of the feed-back table

When the retrocausal effect is missing, the differences of the mean values of the heart rate tend to zero and the lines vary around the baseline (the 0.00 line), whereas the stronger is the effect and the more the lines separate from the baseline.

Using for each subject this graphical representation we discovered that it was necessary to use computer monitors with brilliant colors, it was necessary to conduct the experiment in calm and silent environments leaving the subject alone (except for the first trial which was considered a training trial, and the experimenter was in the room to check if the subject had understood the instructions). Another important consideration was that the attention of the subject had to be on the last two phases; in fact, if little attention was given to the first phase, the effect was still showing. The effect spread all over phase 1, and was not associated only to the color in phase 1 that matches the target color.

Feed-back tables became the row-data for the analyses, which were conducted using non-parametric statistics such as Chi-Square and Fisher's exact test.

Anticipatory learning effect

While performing the first three experiments we came across Antonio Damasio[14] and Antoine Bechara's[15] works and experiments on anticipatory reactions. Studying neurological patients affected by decision-making deficits, Damasio discovered that feelings play an important role allowing to operate advantageous choices, without having to produce advantageous assessments, and that decision-making deficits are always accompanied by alterations in the

---

[14] Damasio, A.R., *Descarte's Error. Emotion, Reason, and the Human Brain*, Putnam Publishing, 1994:
https://www.amazon.it/dp/B00AFY2XVK
[15] Bechara, A., Damasio, H., Tranel, D. and Damasio, A.R., *Deciding Advantageously before Knowing the Advantageous Strategy*, Science, 1997 (275): 1293:
www.labsi.org/cognitive/Becharaetal1997.pdf

ability to feel. Damasio noticed that the absence of feelings leads to the inability to *"feel the future"* and choose advantageously and suggested that goal-oriented systems, moved by finalities, are based on feelings. These systems use body signals coming from the autonomic nervous system: heart, lungs and intestine. Classical measurements of the autonomic nervous system parameters are: heart rate frequency, skin conductance and body temperature.

Bechara, a student following a specialization course in Damasio's laboratory, devised a guessing task in order to test Damasio's hypothesis[16]. In the experiment the subject is seated in front of a table on which 4 decks of cards are placed, each marked with a different letter: A, B, C and D. Subjects receive 2,000 dollars (false, but perfectly resembling true money) and are told that the aim of the game is to lose the least and try to win as much as possible. The game consists in uncovering cards, one at a time, from any of the decks, until the experimenter stops the game. Each card is associated with a gain or a loss of money. Only when a card is turned it is possible to know how much the subject has earned or lost. Subjects start testing each of the decks, searching for clues and regularities. Decks A and B give high gains, but lead to higher losses, while decks C and D give lower

---

[16] Bechara, A., Damasio, H., Tranel, D. and Damasio, A.R., *The Iowa Gambling Task and the somatic marker hypothesis: some questions and answers*, Trends in Cognitive Sciences, 9: 4, April 2005, web.stanford.edu/~jlmcc/papers/BecharaEtAl05_TiCS.pdf

gains, but lead to a slow gain of money. Players gradually develop the knowledge that decks A and B are more dangerous.

Both normal subjects and patients produce skin conductance reactions each time they receive a gain or a loss after they turn a card.

However, in normal subjects, after they have turned a certain number of cards, something different happens. Just before they choose a card from a dangerous deck (A or B) a skin conductance response is observed which increases while the game progresses. Damasio interpreted this as a learning effect. The subject gradually learns the possible negative outcome of each deck, and before a card is chosen the autonomic nervous system informs the subject through the activation of feelings, which in this case were measured using skin conductance.

Patients with decision-making deficit do not show this anticipatory arousal of skin conductance and chose disastrously.

Learning versus retrocausal

The experimental design of the fourth experiment was therefore changed, in order to allow to distinguish anticipatory effects due to learning from anticipatory effects due to retrocausality.

In order to allow for a learning effect, the fourth experiment used different probabilities in the selection of target colors. In the third phase one color had a 35% chance of being selected (lucky color), one had a 15% chance (unlucky color) and the last two colors had a 25% chance (neutral colors).

- Differences in heart rate frequencies observed in phase 1, in association with the unpredictable random selection of the target operated by the computer in phase 3 were attributed to a retrocausal effect, since future selections of the target are unpredictable.
- Differences in heart rate frequencies observed in phase 1, in association with the choice operated by the subject in phase 2, were interpreted as learning effect.

The task given to the subjects was to guess the highest number of target colors. Subjects were not informed that colors had a different probability of being selected and the experimenter did not know which were the lucky, unlucky and neutral colors.

Results

The feed-back tables on the differences between mean values of the heart rate (HR) when the color is target and non-target provided the row data for the statistical analyses of the retrocausal effect. The sample was of 30 subjects. Feed-back tables were divided into 3 groups: first 33 trials (starting from the second trial), central 33 trials and the last 33 trials. For the retrocausal effect the mean values differences were therefore 5760 (30 subject x 16 heart rates x 4 colors x 3 groups of feed-back tables). Using this data-set and a threshold of 1.5, the following distribution was obtained:

Frequencies	Differences of the mean values			Total
	Up to -1.500	-1.499 to +1.499	+1.500 and over	
Observed	1053 (17.83%)	3680 (63.89%)	1027 (18.28%)	5760 (100%)

Distribution of mean HR differences in feed-back tables of the retrocausal effect

In order to assess the effect, we needed a term of comparison which the Chi Square names *expected frequencies* in the absence of an effect. In order to calculate this distribution a "Non-Correlated Target" (NCT) was used, a sequence which in the first trial was blue, then green, then red, then yellow and went on repeating regularly until the 100th trial. This sequence was non-correlated with the random sequence of targets which was selected by the computer and shown to the experimental subject. It was therefore used to calculate the expected frequencies. The following table was obtained:

Frequencies	Differences of the mean values			Total
	Up to -1.500	-1.499 to +1.499	+1.500 and over	
Observed	1053 (17.83%)	3680 (63.89%)	1027 (18.28%)	5760 (100%)
Expected (NCT)	781 (13.56%)	4225 (73.35%)	754 (13.09%)	5760 (100%)

Distribution of mean HR differences in feed-back tables of the retrocausal effect, for observed and expected frequencies

The Chi Square value for this table is 263.86 which is by far more than 13.81 which (with 2 df, i.e. two degrees of freedom) corresponds to p<0.001. In this example it was not possible to use the exact test of Fisher since this test requires 2x2 tables.

The same procedure was used for the learning effect. The mean values differences were 4320 (30 subject x 16 heart rates x 3 types of colors x 3 groups of feed-back tables). Using this data-set and a threshold of 1.5 the following distribution was obtained:

Differences	Color chosen by the subject			Total	NCT
	Neutral	Lucky	Unlucky		
From + 1.5	14.0%	16.6%	17.2%	16.0%	13.1%
- 1.49 to +1.49	73.5%	66.0%	66.0%	68.5%	73.3%
Up to -1.5	12.5%	17.4%	16.8%	15.5%	13.6%
	100% (n=1,440)	100% (n=1,440)	100% (n=1,440)	100% (n=4,320)	100.0%

Distribution of mean HR differences in feed-back tables of the leaning effect

We see that before the choice of neutral colors, observed and expected frequencies coincide (73.5%

compared to 73.3 expected according to NCT), whereas for the lucky and unlucky colors it is possible to observe a difference between observed and expected frequencies which is associated with a Chi Square value of 39.15 ($p<1/10^9$), which shows the existence of a learning effect.

According to Damasio during the experiment the subject learns the different probabilities, and this shows in the form of a stronger activation of the learning effect.

Differences (absolute values)	Trial			Total	NCT
	2-34	35-67	68-100		
Below 1.5	69.4%	73.8%	62.3%	68.5%	73.3%
From 1.5	30.6%	26.2%	37.7%	31.5%	26.7%
	100% (n=1,440)	100% (n=1,440)	100% (n=1,440)	100% (n=4,320)	100.0%

Learning effect. Distribution of mean differences of HR measured in phase 1 according to the choice operated by the subject in phase 2, divided for groups of trials.

The computer selects which are the lucky, unlucky and neutral colors at the beginning of the experiment, using a random procedure. No one during the experiment knew which were the lucky and unlucky colors, only at the end of the experiment this information was saved in the data file and could be known. The hypothesis was that the effect should increase while the experiment progresses and that it would be particularly strong in the last trials.

However, the table shows in the first 33 trials a slight learning effect with a Chi Square value of 11,53, just over 1/1000 of statistical significance. In the middle 33 trials no learning effect is observed. In the last 33 trials the learning effect is strong with a Chi Square of 89,77 which corresponds to $p<1/10^{22}$. These results show a learning effect that is not gradual in its increase, as expected by Damasio.

In the retrocausal distribution of the effect we see a strong effect in the first 33 trials, which drops in the middle trials and then increases again in the last trials.

Differences (absolute values)	Trial			Total	NCT
	2-34	35-67	68-100		
Below 1.5	59.6%	70.8%	61.2%	63.9%	73,3%
From 1.5	40.4%	29.2%	38.8%	36.1%	26,7%
	100% (n=1,920)	100% (n=1,920)	100% (n=1,920)	100% (n=5,760)	100,0%

Retrocausal effect. Distribution of mean differences of HR measured in phase 1 according to the selection operated by the computer in phase 3, divided for group of trials.

If we graphically compare the leaning and retrocausal effect, we have the following graph:

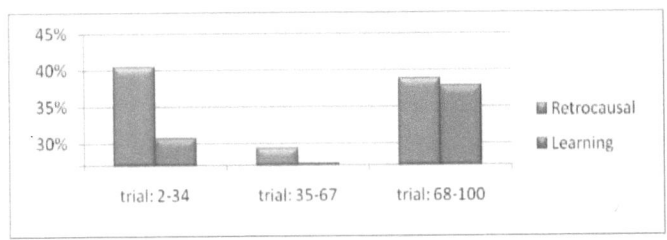

The retrocausal effect is strong starting from the first 33 trials, while the learning effect is just slightly significant. Then, in the middle trials the learning and retrocausal effects drop and disappear. At the end of the experiment, in the last 33 trials, both the effects become strong.

This trend suggests that when the learning effect starts emerging it conflicts with the retrocausal effect, since they both use similar signals coming from the autonomic nervous system, and consequently they both drop. In the last part of the experiment the subject becomes capable of distinguishing between these two similar signals, and thus both the effects emerge again and show strongly.

It was interesting to note that whilst strong learning and retrocausal effects are observed especially in the last 33 trials, this does not translate into more advantageous guesses and subjects continued to guess randomly.

The results have been here described in a non-technical way. A more detailed analysis is available in the books *"Retrocausality: experiments and theory"*, *"A syntropic model of consciousness"* and in *"The methodology of concomitant variations"*[17] which provides the access to the sintropia.ds software which among its tutorials provides the whole original dataset of this experiment.

---

[17] Di Corpo, U. and Vannini, A., *The methodology of concomitant variations,* ISBN: 9781520326634, https://www.amazon.com/dp/1520326637

Comments

The results of these experiments show that syntropy acts mainly on the autonomic nervous system. Since syntropy propagates backwards in time, feelings of warmth and emptiness help us feel the future and orient our choices towards advantageous goals. The following examples provide some insights into the implications of this backward in time flow:

— The article "*In Battle, Hunches Prove to be Valuable*", published on the front page of the New York Times on July 28, 2009, describes how premonitions helped soldiers save themselves: "*My body suddenly became cold; you know, that feeling of danger, and I started screaming no-no!*" According to syntropy, the attack happens, the soldier experiences fear and death and these feelings of distress propagate backward in time. The soldier in the past feels these as premonitions and is driven to take a different decision, thus avoiding the attack and death. According to the New York Times article, these premonitions have saved more lives than the billions of dollars spent on intelligence.

— William Cox, conducted studies on the number of tickets sold in the United States for commuter trains between 1950 and 1955 and found that in the 28 cases where commuter trains had accidents, fewer tickets were sold[18]. Data analysis was repeated verifying all possible intervening variables, such as bad weather conditions, departure times, day of the week, etc. But no intervening variable was able to explain the correlation between reduced ticket sales and accidents. The reduction of passengers on trains that have accidents is strong, not only from a statistical point of view, but also from a quantitative point of view. According to syntropy, Cox's discoveries can be explained in this way: when people are involved in accidents, the feelings of pain and fear propagate backward in time and can be felt in the past in the form of presentiments and premonitions, which can lead to the decision not to travel. This propagation of feelings can therefore change the past. In other words, a negative event occurs in the future and informs us in the past, through our inner feelings. Listening to these feelings can help us decide differently and avoid pain and suffering in our future. If we listen to our inner voice, the future can change for the better.

– Among many possible examples: on May 22, 2010 an Air India Express Boeing 737-800 flying between Dubai and Mangalore crashed during landing, killing 158 passengers, only eight survived

---

[18] Cox, W.E., *Precognition: An analysis.* Journal of the American Society for Psychical Research, 1956(50): 99-109.

the accident. Nine passengers, after check-in, felt sick and could not get on board.

People commonly refer to the heart or to the solar plexus and not to the autonomic nervous system.

Syntropy nourishes the vital functions and is a converging energy that propagates from the future, consequently when the inflow of syntropy is good we feel warmth (ie energy concentration) and well-being in the thoracic area of the autonomic nervous system.

On the contrary, when the inflow is insufficient we feel emptiness, pain and anxiety.

These feelings work like the needle of a compass which points towards the source of syntropy (ie life energy).

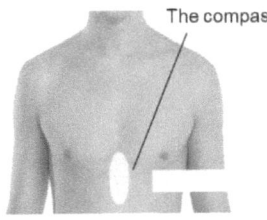
The compass of the heart

The *Attractor*

Unfortunately most people are unaware of how the compass of the heart works and their main concern is to avoid suffering and the unbearable feelings of anxiety. This explains, for example, the mechanism of drug addiction.

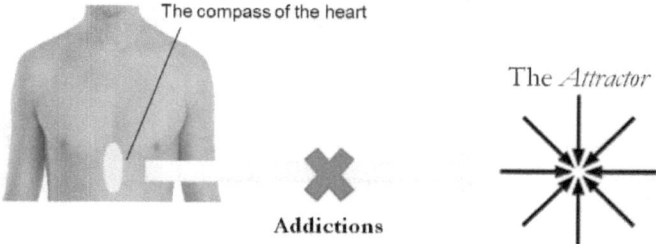

The compass of the heart points to the source of syntropy, but drugs, alcohol and whatever we use to sedate our suffering reduces our possibility to use this compass.

HOW TO ENHANCE PRECOGNITIONS, INTUITIONS AND WEALTH

A very important example of precognition has been provided by Steve Jobs, the founder of Apple Computer. Jobs was able to intuit the future and this was at the basis of his creativity and success.

Jobs had been abandoned by his natural parents and it was the drama that accompanied him throughout his life. He was tormented and never accepted that he had been abandoned.

He left university during the first year and ventured to India to find his inner self. He discovered a completely different vision of the world that marked his change: *"in the Indian countryside people do not let themselves be guided by rationality, as we do, but by intuitions."*

He discovered that intuitions let you perceive the future. A very powerful faculty, very developed in India, but practically unknown in the West.

He returned to the United States convinced that intuitions were more powerful than the intellect. To enhance intuitions he discovered that it was necessary to live a minimalist life, reducing entropy as much as possible. It was important to avoid meat, he became a vegan. It was important to avoid alcohol, tobacco, coffee and any substances that had an effect on the autonomic nervous system impairing the feelings of the "heart". He discovered that it was necessary to calm the chatter of the mind and for this end he practiced Zen meditation. Jobs had the courage to follow his heart and not to be influenced by the judgment of others.

He always tried to reduce entropy to the point that it took him more than 8 months to choose the washing machine. He absolutely had to find the one with the lowest energy consumption and the maximum efficiency. He lived in a thrifty way, a life so essential and austere that led his children to believe he was a poor man.

The way he lived was the result of his need to focus on the heart, on the inner feelings of the autonomic nervous system. He avoided wealth because it could distract him from the voice of the heart. He was one of the richest men on the planet, but he lived like a poor man!

From a syntropic perspective, his minimalist choices allowed intuitions and precognitions to emerge, becoming the source of his revolutionary innovations and wealth.

Jobs opposed marketing studies, as he believed that people usually don't know the future and that only intuitive people can feel the future.

When he returned from India he saw an electronic board at his friend Steve Wozniak's house and he had the vision of a computer that could be held in one hand. He asked Wozniak to develop a prototype of a personal computer, which he named Apple I. He managed to sell a few hundreds of them and this sudden success gave him the courage to develop a more advanced model, suitable for ordinary people, which he named Apple II.

Jobs was not an engineer, he had no scientific or technical mind, he was simply an artist! What do computers have to do with his life? Jobs had nothing to do with electronics, but his intuitive abilities showed him an object of the future. Thirty years earlier, in 1977, he had the vision of the smart phone, a pocket computer that combines aesthetics, simplicity, technology and minimalism! He felt the need for a product that, in addition to being technologically perfect, had to be also beautiful and simple!

His obsession with beauty and simplicity led him to devote an enormous amount of time to the details of the Apple II. It had to be beautiful, silent and at the same time essential and simple! It was an unprecedented commercial success that made Apple Computer one of the leading global companies.

Jobs noticed that when the heart gave him an intuition, it was for him a command that he had to follow, regardless of the opinions of others. The only thing that mattered was finding a way to give shape to the intuition.

For Jobs, the vegan diet, Zen meditation, a life immersed in nature, abstention from alcohol and coffee were necessary to nourish his inner voice, the voice of his heart and strengthen his ability to intuit the future.

At the same time, this caused great difficulties. He was sensitive, intuitive, irrational and nervous. He was aware of the limitations that his irrationality caused in handling a large company, such as Apple Computer, and chose a rationalist manager to run the company: John Sculley, a famous manager he admired but with whom he entered continually in conflict, to the point that in 1985 the board of directors decided to fire Jobs from Apple, the company he had founded.

Apple Computer continued to make money for a while with the products designed by Jobs, but after a few years the decline began and in the mid-1990s it came to the brink of bankruptcy. On December 21, 1996, the board of directors asked Jobs to return as the president's personal advisor. Jobs accepted. He asked for a salary of one dollar a year in exchange for the guarantee that his insights, even if crazy, were accepted unconditionally. In a few months he revolutionized the products and on September 16, 1997 he became interim CEO.

Apple Computer resurrected in less than a year. How did he manage?

He believed that we should not let the noise of others' opinions dull our inner voice. And, more importantly, he repeated that we must always have the courage to believe in our heart and in our intuitions, because they already know the future and know where we need to go. For Jobs, everything else was secondary.

Being *i*nterim marked all his new products. This is the reason why their name was preceded by the letter *i*: *i*Pod, *i*Pad, *i*Phone and *i*Mac.

Jobs' children believed he was poor. They often asked him: "*Daddy, why don't you take us to one of your rich friends?*"

He talked about important business walking in parks or in nature. To celebrate a success, he invited employees to restaurants for $10 per person. When he made a gift he collected flowers in a field. He wore the same clothes for years. Despite the immense wealth he had!

He was convinced that money was not his, but a tool to reach the end. For him money was exclusively a tool.

At the time of Apple I, he repeated that his mission was to develop a computer that could be held in one hand and not to become rich.

The ability to feel the future was the source of Jobs' wealth. It was the ingredient of his creativity, genius and innovation.

Zen meditation helped Jobs calm his mind and focus the attention into the heart.

In his lectures he used to say that almost everything, expectations, pride and fears of failure, vanish in the face of death. He emphasized the centrality of death and the fact that when we are aware of dying we pay attention only to what is really important. Being constantly aware that we are destined to die is one of the most effective ways to understand what is really important and to avoid the trap of attaching ourselves to materiality and appearances. We are already naked in the face of death. Since we must die, there is no reason not to follow our heart and do what we have to do.

Jobs believed in the invisible and in synchronicities. He built the headquarters of Pixar (one of his companies) around a large square where everyone had to go through or stop if they wanted to eat something or use the services. In this way the invisible world was favored by chance encounters.

According to Jobs, chance does not exist. Chance encounters allow the invisible, to activate intuitions, creativity and synchronicities and make visible what is not yet visible.

Jobs loved to quote Michelangelo's famous phrase:

"In each block of marble I see a statue as if it were in front of me, shaped and perfect in attitude and action. I just have to remove the rough walls that imprison the beautiful appearance to reveal it to others as my eyes see it."

Jobs believed that we all have a task, a mission to carry out. We just need to discover this mission by removing what is not necessary.

Jobs made visible what he had intuited. He died a few months after the presentation of the *i*Phone, the computer that can be held in one hand, the mission of his life.

Jobs' life testifies that intelligence and creativity come from the future, from the invisible and that we can access the invisible through intuitions.

He showed that the voice of the heart brings the future into the present.

Rainer Maria Rilke said: *"The future enters us, to become us, long before it happens."*

An important mechanism that enhances syntropy is provided by complementarity. Entropy and syntropy can be represented as a seesaw with causality on one side and retrocausality on the other side.

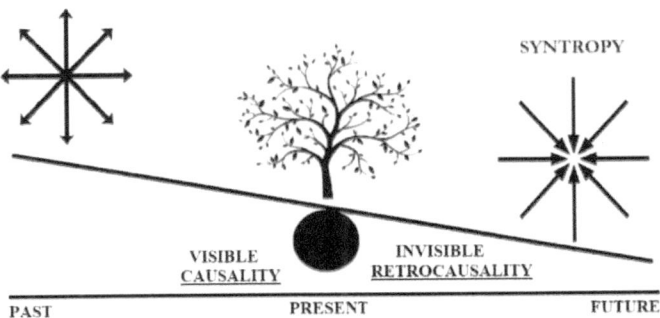

Life is the manifestation on the physical plane of syntropy an retrocausality. It is constantly in conflict with entropy and must always diminish it. However, this is hampered by our activities that tend to increase entropy.

The challenge we constantly face is to *increase syntropy, reducing entropy and remaining active.* Most people are not aware of this challenge and tend to live in dysfunctional ways that increase entropy and reduce syntropy.

To describe this challenge I will use the example of a freelance, single, whose expenses exceeded the income of over five hundred euros a month.

The savings were running out and he had no one to ask for help. He started reducing his expenses: no money in his wallet, no credit on his cell phone. But things went from bad to worse. At this point he asked me for help.

Let's see how it went:

«How much do you spend on your mobile phone?»
«About 40 euros a month, but I always find myself without credit.»
«Why don't you change provider? There are interesting promotions. With only 10 euros a month you can have unlimited minutes and SMS and 20 gigabytes of internet.»

Lowering entropy means saving, but this must be done by maintaining or increasing the quality of life. For example, by changing an old contract. In this case, changing provider and choosing a new contract has led to an increase in the quality of life and to save over three hundred euros a year!

The trick is to improve the quality of life by saving.

When entropy (expenses) and syntropy (incomes) are balanced, the invisible world begins to manifest.

In this example we need to reduce spending by at

least six thousand euros a year.

> *«Do you take shirts to the laundry to be ironed?»*
> *«I wash them, but I am not able to iron them. I take them to the laundry to have them ironed.»*
> *«How much does it cost you?»*
> *«Between 50 and 70 euros a month.»*
> *«Why don't you ask your maid if she can iron them for 8 euros more per month?»*

The maid immediately accepted. Another small optimization that led to save over six hundred euros a year, but which significantly increased the quality of life by eliminating the hassle of going to the laundry. Again an increase in the quality of life while saving! These first two optimizations reduced entropy by around one thousand euros a year and increased the quality of life. The goal is to reach six thousand euros to balance income and expenses.

> *«Do you go to work by car?»*
> *«I also use the scooter to save money, but the traffic is really dangerous!»*
> *«Why don't you use your bicycle?»*
> *«On these roads ?!»*
> *«No, on alternative roads.»*
> *«My house is in the city center, the office is not far away, but I have always considered the bicycle impossible due to the difference in altitude of over 30 meters. I would arrive tired and sweaty.»*

«If you have to climb it is better to choose a steep but short road, get off and push, rather than pedaling.»

Thus he discovered the beauty of the streets of the city center and parks. In less than 25 minutes he could reach his office by bicycle. It took more time by car or scooter. The next day he sold the scooter, canceled the insurance and the garage. Another three thousand euros saved per year. With this simple optimization, he received other advantages: he exercises and no longer needs to go to the gym, more money and time saved! Moreover, he spends less on fuel.

Entropy has now decreased by over four thousand euros a year and the quality of life has improved!

We need to find another two thousand euros before syntropy, the invisible world can begin to show.

«Your electricity bill exceeds 200 euros every two months! As a single you should not pay more than 50 euros.»
«What should I do?»
«Try using low energy light bulbs, such as LED lamps, and set the timer to the water heater.»

Small changes that required little time and money. One hundred and fifty euros saved every two months, nine hundred euros a year. With this small optimization he felt consistent with his ecological beliefs and the quality of life increased. Now he had reduced his expenses by over five thousand euros a year! We must reach the goal of six thousand euros a year!

«How much do you pay for electricity in your office?»
«About 300 euros every two months.»
«Do you use halogen bulbs !?»
«Yes.»

He discovered that he could save over a thousand euros a year simply by replacing the halogen spotlights with LED spotlights.

Now that the expenses no longer exceed the incomes, syntropy can begin to show in the form of synchronicities: meaningful coincidences.

Jung and Pauli have coined the term synchronicity to indicate an invisible causality different from that familiar to us. Synchronicities manifest as meaningful coincidences, because they converge towards an end.

In the psychological literature of the 20th century Carl Gustav Jung and Wolfgang Pauli added synchronicities (syntropy) to causality (entropy). According to Jung, synchronicities are the experience of two or more events that are apparently causally unrelated or unlikely to occur together by chance, yet they are experienced as occurring together in a meaningful manner.

The concept of synchronicity was first described in this terminology by Carl Gustav Jung in the 1920s. The concept does not question, or compete with, the notion of causality. Instead, it maintains that just as events may be grouped by causes, they may also be grouped by finalities, a meaningful principle. Jung coined the word synchronicities to describe what he called *"temporally coincident occurrences of acausal events."* He variously described synchronicities as an *"acausal connecting principle," "meaningful coincidence"* and *"acausal parallelism."*

Jung gave a full statement of this concept in 1951 when he published the paper *Synchronicity - An Acausal Connecting Principle*,[19] jointly with a related study by the physicist Wolfgang Pauli.

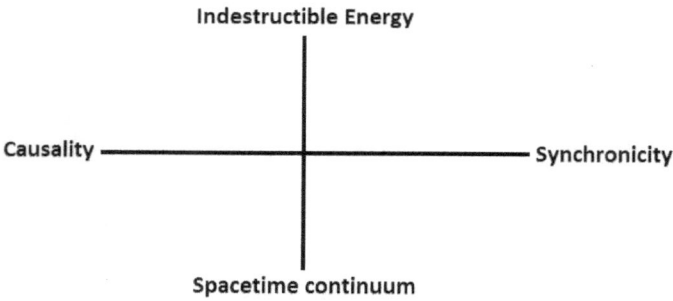

In Jung's and Pauli's description causality acts from the past, whereas synchronicities act from the future. Synchronicities are meaningful since they lead towards a finality, providing a direction to events which correlate them in an apparently acausal ways.

Jung and Pauli described causality and

---

[19] Jung C.G. (1951), *Synchronicity - An Acausal Connecting Principle*, Princeton University Press, www.amazon.com/Synchronicity-Connecting-Principle-Collected-Bollingen/dp/0691150508

synchronicities acting on the same indestructible energy. They are united by this energy, but at the same time they are complementary.

Synchronicities act from the future and group events according to purposes and are significant because they have purpose.

«How much do you pay for renting your office?»
«Nothing. It is owned by my aunts.»
«They could rent it and make a profit, but you use it for free?!»
«Exactly.»
«And what are your aunts living on?»
«They both receive a pension and have some savings, but their financial situation is not good, they constantly complain.»
«Have you ever thought about renting a room in an office and letting your aunts rent their apartment?»
«I have no money, I can't afford to pay a rent!»
«How's your business going?»
«I have few clients, perhaps because of the economic crisis, but also because of the position of the office.»
«A less prestigious office, but in a strategic and well-connected place could help you have more customers?!»

The first synchronicity is the following. The day after this dialogue, as if by magic, he received the offer of a room in an office in the most central area of the city, at the price of only 250 euros a month, including all utilities! The aunts' apartment was in a very beautiful and prestigious place, but difficult to reach and there was no parking: beautiful, prestigious, but inconvenient and very expensive. However he hesitated, he didn't dare!

The next day the second synchronicity occurred. He received a call from an airline company that offered 2,800 euros a month for his aunts' apartment. Obviously the aunts asked him to find another place immediately and fortunately the day before he had received the offer of a room. But he still wasn't convinced. The office in the city center was in a very noisy area: well connected, but chaotic.

The third synchronicity is the following. That same afternoon he was walking in the area of the city he likes most. It is not central, but it is green, quiet and well connected. At a shoemaker's window, he saw a notice for a room in an office. The apartment was in the building next to the shoemaker. He called and immediately went to see it. He instantly decided to rent the room. In a city like Rome it is difficult to find rooms for rent in professional studios and above all in such a beautiful place of the city.

When synchronicities are activated, we are attracted to places and situations that otherwise we would not have taken into consideration and that solve our problems. Synchronicities are accompanied by feelings of warmth and well-being in the thoracic area that inform us that we are on the right path.

«*I began to feel warmth and well-being in the chest area. My clients like the new studio. There is a parking lot, it is nice, quiet and it is located near a metro station. My business is thriving, my savings are increasing and my personal and sentimental life has improved.»*

Syntropy offers wealth and happiness. But when things go well it is easy to fall back into the old entropic and dissipative lifestyles.

A few months later he received a job offer, a prestigious job abroad: his dream!

He immediately accepted and moved. The salary was high, taxation was low. Suddenly he would become a rich man who could lead the rich life he had always wanted.

But this reverses the balance between entropy and syntropy: wealth leads to living in an entropic way, entropy increases and syntropy decreases and we go back to failure!

«The foreign company was only interested in making money, without any ethics. I had to work almost fifty hours a week, there was nothing else outside the company. It was necessary to give absolute priority to what was profitable, even if immoral. A few months later I felt disgusted with my profession. Taxes were low, but I had to pay all the services. By adding the rent of the house and the expenses related to the fact that I was a foreigner, I paid much more than I earned. After only six months I had accumulated more than twenty-eight thousand euros of debts! The dream had shattered and had become a nightmare. From heaven I fell

to hell. I had no time for myself or for my love life. First I felt discomfort, then suffering, and eventually depression and anxiety exploded. I decided to go back to Italy!»

This often happens. Syntropy increases the quality of life, well-being, but also wealth. As soon as material wealth returns people fall into entropic lifestyles.

For this reason the increase in syntropy must be accompanied by an inner transformation. People do not have to consider money as their property, but as a tool. They must be aware that happiness is not achieved through wealth, but thanks to the fulfillment of our mission.

If this inner transformation is lacking, the process fails.

Wealth is only one aspect of the game between entropy and syntropy. When wealth is reached without an inner transformation it is inevitable to fall back into entropy and suffering.

The compass of the heart is of great importance in the game of life, but since in the same area we perceive emotions linked to fear and danger it is not easy to use it. These emotions are activated by the amygdala. The amygdala is designed to ensure survival.

When we are faced with a danger it releases hormones that trigger the fight or flee reaction. The amygdala is fast, but inflexible. The emotional charge enters our body and covers the feelings of the heart.

Fears and dangers limit the ability to use the compass of the heart and they increase entropy.

The compass of the heart requires that we silence fear and the chatter of the mind. A very effective way is provided by Zen meditation.

During Zen meditation participants cannot react to stimuli, but they can only observe them. Practicing Zen meditation we discover that thoughts wait for the reaction of the heart. When the heart reacts it provides energy to the thought which becomes stronger. When we don't react the thought dissolves.

The heart decides when to react and to be silent; the mind can only adjust to the will of the heart. We are the heart. Our will is in the heart. In this way the scepter of command moves from the head to the heart and the mind becomes silent.

Silence is a natural technique, a simple and enjoyable way of being together with others. It is not a religion and does not require devotion to a faith, or to a specific philosophy. It creates distance from our thoughts. It frees our being from the conditioning power of the words and leads to discover that we are part of something broader. When the chatter of the mind ends we experience a new condition: to be without thinking. A state in which thoughts are produced only when required by the heart. A state in which the gap between a thought and the other is not empty, but it is pure and absolute potentiality. Being without thinking empowers the heart, intuitions and precognitions.

Another factor which influences the perception of the heart is what we eat.

John Hubert Brocklesby became a vegetarian in prison during the First World War.[20] For him, Christians did not have to kill other Christians and declared himself a conscientious objector. He was arrested and imprisoned in the Richmond Castle. He had to face court martial. He knew he would be sentenced to death and he was terrified at the idea.

Another conscientious objector told him: «*If you talk with your heart, God speaks through you.*» This gave him courage and he started saying in front the court martial what he felt in his heart. Then this same conscientious objector added: «*If you do not eat meat, the voice of the heart becomes stronger.*»

John Hubert Brocklesby became a vegetarian in prison to serve the will of God and face court martial.

Since we have a vegetarian structure (no claws to hunt, our teeth are suitable for fruit and the digestive system is long, not meant for meat) the attractor towards which we are evolving has these features. Therefore, being vegetarian helps the connection with the future, increasing the flow of syntropy, intuitions and precognitions.

Among the diet options that seem to increase the

---

[20] Jones WE, *We Will Not Fight: The Untold Story of World War Ones Conscientious Objectors*, www.amazon.com/dp/1845133005/

perception of the heart one is liquidarism.

Michael Werner, born in 1949 in northern Germany and CEO of a pharmaceutical research institute in Arlesheim, became liquidarian in January 2001 and since then drinks only water and does not eat solid food. In his book *Living on Light* Werner says that:

> *"I found that my conversion to living without food went extraordinarily well. I expected to feel weaker and weaker during the first few days. But then I began to realize that in my case this weakness did not exist. Instead I experienced a growing feeling of lightness during the day and a decrease in the amount of sleep I needed during the night. Going through this process was probably the most intense experience of my adult life."*

If it is true that one can live and be fit and healthy without eating, incredible scenarios open about human life and life in general.

Werner notes that being liquidarian is different from fasting:

> *"It is something completely different! With fasting the body mobilizes reserves of energy and matter and one cannot fast for an unlimited time, nor can one be without drinking. But the process I was undertaking was and remains a mental-spiritual phenomenon that requires a particular inner predisposition. In reality there is a condition: opening up to the idea of being able to be nourished by the etheric, by*

prana or by whatever it may be called. This is the necessary requirement. Then it will happen. I live liquidarism as a gift from the spiritual world."

Rudolf Steiner (1861-1925), an Austrian philosopher, social reformer, architect and esotericist, attempted to formulate a spiritual science, a synthesis between science and spirituality that applied the clarity of scientific thought, of Western philosophy, to the spiritual world. Steiner believed that matter was condensed light (he used the word light with the same meaning of syntropy). If matter is condensed syntropy, there must be ways to transform the invisible (syntropy) into matter. Our visible environment is immersed in an invisible one, a syntropic reality that offers incredible possibilities, including that of living from syntropy. Steiner believed that life was impossible without syntropy (ie without light), since syntropy is the vital energy that we continuously and directly absorb. Living only on water requires that we believe possible to *"live by syntropy."* According to Steiner, the act of digesting stimulates the body to absorb the vital energy from the invisible, which is transformed and condensed into substances that maintain and build our body. Steiner used the following example: when we eat a potato, we chew and digest and this leads to absorbing the vital forces from our etheric environment and condensing them into substances.

Our body acquires structure and substance absorbing syntropy and invisible forces.

Michael Werner emphasizes that the only prerequisite for feeding on light (ie syntropy) is to believe in it. He uses the words of Steiner:

"There is a fundamental essence of our earthly material existence from which all matter is produced through a process of condensation. What is the fundamental substance of our terrestrial existence? Spiritual science gives this answer: every substance on earth is condensed light! There is nothing but condensed light ... Wherever you touch a substance, there you have condensed light. All matter is, in essence, light."

In other words, all matter is nothing else but condensed syntropy!

RETROCAUSALITY: CHANGING THE PAST

We are accustomed to the fact that causes always precede their effects. But, the *energy-momentum-mass* equation predicts three types of time:

- *Causal time*, is expected when the forward-in-time energy solution prevails. That is when systems diverge, such as our expanding universe. In diverging systems entropy prevails, causes always precede effects and time flows forwards, from the past to the future. Since entropy prevails, no advanced effects are possible, such as light waves moving backwards-in-time or radio signals being received before they are broadcasted.
- *Retrocausal time*, is expected when the backward-in-time energy solution prevails. That is when systems converge, such as black-holes. In converging systems retrocausality prevails, effects always precede causes and time flows backwards, from the future to the past. In these systems no delayed effects are possible and this is the reason why no light is emitted by black-holes.
- *Supercausal time* would characterize systems in which diverging and converging forces are balanced. An example is offered by atoms and quantum mechanics. In these systems causality

and retrocausality coexist and time is unitary: past, present and future coexist.

This classification of time recalls the ancient Greek division in: *Kronos*, *Kairos* and *Aion*.

- *Kronos* describes the sequential causal time, which is familiar to us, made of absolute moments which flow from the past to the future.
- *Kairos* describes the retrocausal time. According to Pythagoras kairos is at the basis of intuitions, the ability to feel the future and to choose the most advantageous options.
- *Aion* describes the supercausal time, in which past, present and future coexist. The time of quantum mechanics, of the sub-atomic world.

Since syntropy and entropy coexist at the quantum level, i.e. the *Aion* level, life must originate at this level. This statement is now supported by the fact that the functioning of living systems is widely influenced by quantum events: the length and strength of hydrogen bonds, the transmission of electrical signals in the microtubules, the action of DNA, the folding of proteins.

A question naturally arises: how do the properties of life ascend from the quantum level of matter, the *Aion* level, to the macroscopic level, the Kronos level, transforming inorganic matter into organic matter?

In 1925 the physicist Wolfgang Pauli discovered in water molecules the hydrogen bond. Hydrogen atoms in water molecules share an intermediate position between the sub-atomic level (*Aion*) and the molecular level (*Kronos*), and provide a bridge that allows the properties of syntropy to flow from the quantum to the macro level.

Hydrogen bonds make water different from all other liquids, increasing its attractive forces (syntropy), which are ten times more powerful than the van der Waals forces that hold together other liquids, with behaviors that are in fact symmetrical to those of other liquid molecules.

Consequently, life originates at the quantum level and rapidly grows into the macroscopic level which is governed by the law of entropy.

In order to survive the destructive effects of entropy, life needs to acquire syntropy from the quantum level and water provides the mechanism, becoming in this way vital.

Among the anomalous properties of water which recall the cohesive qualities of syntropy[21]:

- When water freezes it expands and becomes less dense. Other liquid's molecules, when they are cooled concentrate, solidify, become more dense and heavy and sink. With water exactly the

---

[21] Ball P. (1999), *H₂O A Biography of Water*, Phoenix Book, London.

opposite is observed.
- In liquids the process of solidification starts from the bottom, since hot molecules move towards the top, whereas cold molecules move towards the bottom. The liquid in the lower part is therefore the first which reaches the solidification temperature; for this reason liquids solidify starting from the bottom. In the case of water exactly the opposite happens: water solidifies starting from the top.
- Water shows a heat capacity by far greater than other liquids. Water can absorb large quantities of heat, which is then released slowly. The quantity of heat which is necessary to change the temperature of water is by far greater than what it is needed for other liquids.
- When compressed cold water becomes more fluid; in other liquids, viscosity increases with pressure.
- Friction among surfaces of solids is usually high, whereas with ice friction is low and ice surfaces result to be slippery.
- At near to freezing temperatures the surfaces of ice adhere when they come into contact. This mechanism allows snow to compact in snow balls, whereas it is impossible to produce balls of flour, sugar or other solid materials, if no water is used.
- Compared to other liquids, in water the distance between melting and boiling temperatures is very

high. Water molecules have high cohesive properties which increase the temperature which is needed to change water from liquid to gas.

Water is not the only molecule with hydrogen bonds. Also ammonia and fluoride acid form hydrogen bonds and these molecules show anomalous properties similar to water. However, water produces a higher number of hydrogen bonds and this determines the high cohesive properties of water which link molecules in wide dynamic labyrinths.[22] Other molecules that form hydrogen bonds do not reach the point of being able to build networks and broad structures in space. Hydrogen bonds impose structural constraints extremely unusual for a liquid. One example of these structural constraints is provided by crystals of snow. However, when water freezes hydrogen bonds stop working and the flow of syntropy from micro to macro stops, bringing life to death.

Hydrogen bonds make water essential for life: water is ultimately the lymph of life which provides living systems with syntropy. Water is the most important molecule for life, which is necessary for the origin and evolution of any biological structure. Consequently, if life would ever be discovered beyond

---

[22] Bennun A. (2013), *Hydration shell dynamics of proteins and ions couple with the dissipative potential of H-bonds within water*, Syntropy 2013 (2): 328-333.

Earth water would necessarily be present.[23]

The *energy-momentum-mass* equation suggests that the present is the meeting point of causes that act from the past (causality) and attractors that act from the future (retrocausality).

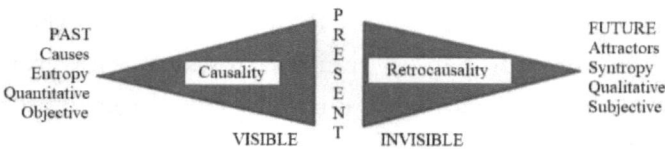

Causality requires a big cause for a great effect. This is due to the fact that causality diverges and tends to disperse. On the contrary with attractors the effect is amplified. The smaller the cause, the more it can be amplified and the greater is the effect.

This strangeness of attractors was discovered in 1963 by the meteorologist Edward Lorenz. When dealing with water, as happens in meteorology, a small variation can produce an amplifying effect. Lorenz described this situation with the famous phrase: *"The flap of a butterfly in the Amazon can cause a hurricane in the United States."*

However, for this to happen it is necessary that the small flap (the active principle) is in line with the attractor. Otherwise entropy prevails and the small

---

[23] Vannini A. (2011) and Di Corpo U., *Extraterrestrial Life, Syntropy and Water*, Journal of Cosmology, journalofcosmology.com/Life101.html#18

energy of the flap is lost. On the contrary, when the variation is in line with the attractor it is amplified.

The hydrogen bond of water operates in both directions: from the micro to the macro, amplifying the effect, and from the macro to the micro informing the attractor. This can provide an insight on how homeopathic remedies work.

Homeopathy is based on water. When we insert into water the similar, the *simillimum*, of what we want to cure, its information enters the quantum level and informs the attractor. The greater the dilution, the greater is the contribution of the attractor in the amplification of the effect.

Homeopathy is the subject of ferocious attacks. In Italy the famous scientific television journalist Piero Angela reiterates that *"homeopathy is fresh water"*, *"pseudoscience"* or even *"magical practice"* and constantly emphasizes that it has no scientific validity: *"It is a placebo effect, this is what the scientific community says."* For Rita Levi Montalcini (Italian Nobel Prize) homeopathy is potentially harmful because it distracts patients from valid treatments and for Renato Dulbecco (another Italian Nobel Prize) it is a practice without any value. Lately the attacks have intensified and the main accusation is that homeopathy is only fresh water.

However, experimental studies show the effectiveness of homeopathy.

Homeopathy was discovered in 1796 by the German doctor Samuel Hahnemann (1755-1843).

This system is based on the so-called law of similes, according to which the remedies must use substances that cause similar symptoms in healthy individuals. These substances are then diluted in water. The strange fact is that the higher is the dilution the more powerful is the effect. The most powerful remedies are those in which the substances have been diluted to the point that it is impossible for a single molecule to still be present in the remedy. For conventional medicine, after removing the active ingredient through dilution, effects can only be placebo effects, not attributable to the remedy, since no solid molecule of the active ingredient is present.

Syntropy claims that the active ingredient, when placed in water, creates links with attractors. So by removing the active ingredient through dilution, these retrocausal bonds remain and are no longer related to the substance but are free to act on any other structure.

Syntropy explains the effects of homeopathy as a consequence of the retrocausal properties of water.[24] The remedies act from the future and the effects are the result of the interaction between causality that is governed by entropy and retrocausality that is governed by syntropy.

When using a substance that induces in the future of a healthy person symptoms similar to those

---

[24] Paolella M.., *Homeopathic Medicine and Syntropy*: http://www.sintropia.it/journal/english/2014-eng-2-01.pdf

observed in a sick person and this substance is diluted in water (beyond the value of Avogadro), the future begins to retroact into the present.

With causality in order to increase the effect it is necessary to increase the cause (the active substance), while with retrocausality in order to increase the effect it is necessary to reduce the cause. Retrocausality works in a way symmetrical to causality. This explains why in homeopathy in order to enhance the remedy instead of increasing the active substance it must be diluted.

Homeopathy cannot be explained on the basis of classical causality, since the active ingredient is completely removed from homeopathic preparations (which are water based). The therapeutic effects, however, are strong and can be proved experimentally. The effects are strong even when no placebo effect is possible, as in the case of studies carried out on plants in agriculture.

The retrocausal properties of water are due to the hydrogen bond. The hydrogen atoms are in an intermediate position between the subatomic (quantum) and the molecular level and provide a bridge that allows syntropy to flow from the attractor to the macroscopic level.

Now let's see two totally different examples of retrocausality.

The first one dates back to 2012. With Antonella I

was hosted by one of the most powerful healers in the United States. We were there to attend a SAND conference (Science and Non-Duality). In the same days the baseball final was held in San Francisco, and the San Francisco Giants were the worst team. Our friend tried to help the Giants by healing them using a technique which worked remotely. However, the effects were scarce, difficult to assess. We told him that according to syntropy, the results could be enhanced thanks to the butterfly effect, using a retrocausal procedure.

The procedure that we suggested was to record the game, without looking at it or knowing the result, then, at the end of the game, our friend had to watch the recording and start using his healing techniques on the players, on a game that had already ended. It might have been a coincidence, but as soon as he started using this retrocausal procedure, the Giants started winning, obtaining increasingly surprising results and managing to achieve what no other team had previously achieved in the history of American baseball. The results were so incredible that they motivated us to continue, and we went on getting always more amazing results. There is a short video made in San Francisco with our healer friend. The link is: youtu.be/ubdNpH-zPwo.

Another example taken from political fiction provides a good idea of what the power of retrocausality could be.

We continuously hear about Climate Changes. There is no doubt that CO_2 is rising. But if we look at it from a broader perspective, the picture seems quite different. In this regard, the past can tell us a lot about the future.

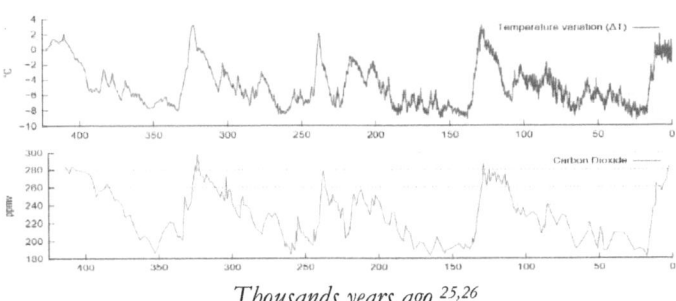

Thousands years ago [25,26]

Ice core data on carbon dioxide (CO_2) and temperatures, available for the last 800 thousand years, show that our planet undergoes warm periods of approximately 10 thousand years and cold ice age periods of about 100 thousand years. CO_2 is produced by life activities such as breathing and decomposition and increases during the warm periods.

In 1972 the chairman of the conference *The Present*

25
en.wikipedia.org/wiki/Ice_age#/media/File:Vostok_Petit_data.svg
[26] CDIAC – Carbon Dioxide Information Analysis Center
http://cdiac.ornl.gov/images/air_bubbles_historical.jpg
http://cdiac.ornl.gov/trends/co2/ice_core_co2.html

Interglacial, How and When Will it End[27] which was held on 26 and 27 January 1972 at the Brown University sent a letter to President Nixon[28] informing that we have already entered the next ice-age: *"we feel obliged to inform you on the results of the scientific conference held here recently...The present rate of the cooling seems fast enough to bring glacial temperatures in about a century, if continuing at the present pace."* In 2015 it was discovered that these fluctuations are caused by a double dynamo effect between two layers of the Sun, one near the surface and one inside its convection area. This model explains the irregularities of the past and predicts what will happen in the future. Valentina Zharkova, one of the discoverers of this double dynamo effect, describes the results in this way[29]: *"We found magnetic waves that appear in pairs, originating from two different layers within the Sun. Both have a cycle of about 11 years, even if they are slightly out of phase. During the cycle, the waves float between the northern and southern hemispheres of the Sun. Combining these waves and comparing them with the real data for the past solar cycles, we found that our predictions are 97% accurate."* Using this model we see that waves will

---

[27] Summary report of the conference *When Will the Present Interglacial End?* Science, 13 Oct 1972, Vol. 178, Issue 4057, pp. 190-202

[28] A copy of the letter is available at: realclimatescience.com/2017/11/the-history-of-the-modern-climate-change-scam/

[29] Royal Astronomical Society – *Irregular heartbeat of the Sun driven by double dynamo* https://www.ras.org.uk/news-and-press/2680-irregular-heartbeat-of-the-sun-driven-by-double-dynamo

become increasingly out of phase during cycle 25, which reaches its peak in 2022. In cycle 26, which covers the decade from 2030 to 2040, they will become totally out of phase and this will cause a significant reduction in solar emissions. *"In cycle 26, the two waves are opposed to each other, with their peak at the same time but in opposite hemispheres of the Sun. Their interference will be destructive and will cancel each other out ... when the waves are in phase, they can show a strong resonance, and we have strong solar activity. When they are out of phase, we have solar minima."* The double dynamo model forecasts a 60% drop in solar activity starting in the 2030-2040 period. When solar emissions decrease, the magnetic shield that protects the planet weakens and cosmic rays enter the core, activating magma and causing strong earthquakes and volcanic eruptions. More than a million volcanoes lie under the sea level against 15,000 on land. Increased eruptions of submarine volcanoes rise ocean temperatures, causing extreme weather conditions such as violent hurricanes and the increase in the amount of water vapor in the atmosphere. There are 3 possible scenarios:

- Humanity will migrate towards the equatorial strip and built towns in the areas which were before covered by the oceans, since these areas are the warmest. But at the end of the ice age the ice caps will melt quickly and give place to interglacial lakes which will suddenly flood the ocean basins wiping away these remains of human civilizations.

At the end only a small portion of humanity might survive and little traces of the previous civilizations and cultures will be left.

- A small elite is concentrating huge resources thanks to the mechanism of private central banks (FED) and public debt, and building advanced shelters which should allow a limited humanity to survive during the one hundred thousand years of glacial temperatures. This scenario is highly entropic and it is doomed to failure.
- Humanity as a whole will be able to face the ice-age and survive, but this requires the shift from the present highly entropic and energy dissipative system to a new system that is syntropic, energy and heat concentrating.

In the evening of November 8, 2016, when the polls closed for the election of the president of the USA, all the gamblers gave Trump's victory a possibility of less than five percent, while Hillary was considered to be certain, with more than ninety-five percent of possibilities. But the impossible turned out to be possible! Trump's election has always remained a mystery! Everyone was sure about Clinton, and since Clinton wanted to start a war against Russia, Russia was accused of having meddled in the elections. All the investigations showed that cheating and manipulation had occurred only on the Clinton side. So what happened?

The FED had prepared the Third World War, it was time to start the scenario number 2. NATO had already surrounded Russia and was ready. The Third World War was planned to start a few days after Hillary's victory. The reason for the war was Putin. Putin had nationalized the Russian Central Bank and was freeing the world from the tyranny of the FED, of the dollar and of public debts. Trump was deeply against wars. Since the beginning of his campaign he declared that he would develop a positive cooperation with Russia, he respected Putin. The scenario was simple: on one hand the Third World War that would have led to the extinction of the majority of humanity, on the other hand Trump, with all his faults, but who would never authorize the war.

Trump was indeed the most unlikely candidate, Clinton herself had helped him win the Republican Nomination, because she was sure she could beat him easily. He was depicted as an irresponsible, an idiot, a danger to the United States and the world.

Trump had the media and the FED against. There was no way for him to win! But Clinton would have brought humanity to extinction and to the second scenario: a small elite of humans who had concentrated incredible resources in order to overcome the Ice Age.

In such a scenario the attractor (syntropy) could not remain neutral, witnessing the extermination of humanity.

In order to act, a healer was necessary, working in

a retrocausal modality. After the ballots closed, this healer had to slightly move his hands bridging the intention of the attractor with the consciousness of people. This tiny ripple, like the flap of the wings of a butterfly, became a hurricane that retrocausally hit American citizens just before casting the vote, awakening their consciences. This minimal ripple turned the impossible into possible. It allowed the victory of Trump, a candidate who had no chance of winning!

People voted for Trump according to their free will, as the Giants played and won with their own forces. The vote was successful. Millions of Americans have voted freely and consistently with their free will. It was a vote against the establishment, against the dictatorship of the FED. Hillary was the candidate of the FED. The media supported Hillary, while it ridiculed Trump, describing him as an incompetent without any experience. The elections were rigged from the beginning! But in favor of Hillary!

Traditionally we think that power is exercised only on the physical and material plane. But Trump proved the opposite. There is another plane, not physical and invisible. No one figured out who was behind Trump, what power was supporting him. The elections had been arranged in favor of Hillary, but Trump won thanks to the butterfly effect! Trump's election avoided the Third World War. A war that the FED had planned to the smallest details. The outcome was supposed to be the extinction of humanity before the advent of the Ice Age.

Trump became the great enemy! They did everything to eliminate him. First they accused him of being an agent of Putin, then they invented sex scandals, then the story of a spy poisoned by the Russians in Britain, then the use of chemical weapons by the Syrian government. They tried several times to eliminate him physically, for example by tampering with the presidential plane to make it explode in flight. But Trump did not trust the CIA and the FBI, he used his own security, and he always managed to avoid the attacks.

They invented all the possible scandals, and brought his entourage to resign and surrounded him with men working for the FED. A coup. An isolated president, but up to the last moment, his position was not to give consent to the Third World War.

This political fiction example suggests that a very small movement, almost imperceptible, but in a calm sea where even the slightest ripple can be amplified infinitely by the attractor, a tiny flap of wings can change the history of humanity. From extinction we have moved to the possibility of the third scenario, a syntropic future, which we must now build together.

EPILOGUE

Generally we tend to overlook the invisible dimension as it is widely believed that it does not exist and that decisions should be based only on facts. This attitude has led people away from retrocausality and intuitions and has limited western cultures only to rational processes that increase entropy.

In the book *"The Voice of Truth"* Gandhi writes:

"There is an indefinable mysterious power that pervades everything. I feel it, although I do not see it. This invisible force makes itself felt and yet challenges any demonstration, because it is so different from everything that I perceive with the senses."[30]

We here suggest the hypothesis[31] of a dimension vital to life, which is invisible to us, although we can feel it in subjective and qualitative ways.

This dimension has been kept intentionally hidden for centuries.

---

[30] Gandhi MK (1968), *The Voice of Truth*, Nvajivan Trust, Ahmedabad.
[31] Di Corpo U and Vannini A (2014), *The balancing role of Entropy / Syntropy in Living and self-organizing systems: QUANTUM PARADIGM*, www.amazon.com/dp/B00KL4SP70

The satirical novella *Flatland*, written in 1884, well describes this situation:[32]

"It is true that we have really in Flatland a Third unrecognized Dimension called 'height', just as it also is true that you have really in Spaceland a Fourth unrecognized Dimension, called by no name at present, but which I will call 'extra-height.' But we can no more take cognizance of our 'height' than you can of your 'extra-height.' (...) Well, that is my fate: and it is as natural for us Flatlanders to lock up a Square for preaching the Third Dimension, as it is for you Spacelanders to lock up a Cube for preaching the Fourth. Alas, how strong a family likeness runs through blind and persecuting humanity in all Dimensions! Points, Lines, Squares, Cubes, Extra-Cubes -- we are all liable to the same errors, all alike the Slavers of our respective Dimensional prejudices."

Who ventures in the realm of Syntropy and Supercausality finds little support, but nevertheless the discovery of this invisible dimension is now sprouting everywhere.

---

[32]Abbott EA (1884), *Flatland*, Seely & Co, UK.

NOTES

www.ingramcontent.com/pod-product-compliance
Lightning Source LLC
Chambersburg PA
CBHW021008180526
45163CB00005B/1937